MOOSE BOT
BOOKS
copyright 2014

Dedicated to Gary and his love of all things science and numbers

Dr. Lou decided to add to his crew

How many bots did it take to have two?

He was still as busy as could be

So he thought "Three would be much better for me"

But then he added two more

And so he actually had four

He built bots to build bots

Multiplying by ten he had lots

How many bots could there be?

Six rows each have ten plus the four groups of three

Soon the bots were all around

And he decided that it was time to cut down

For he had run out of space

In his very small place

He cut them by half and then half once more

It wasn't all done, he had to settle the score

He subtracted five and then subtracted just one

From here he began his mathematical fun

Right now he was left with half of twenty-four

He decided to divide in half once more

Left with no choice but to subtract some again

He said goodbye to a very dear friend

He removed two but then added three

Leaving just six plus me

Can you see the bot by the door and the one by the tree?

Dr. Lou thought about bots no more

Because he's fast asleep on his workshop floor

How many bots did he send away?

I don't know because he won't say

At least not anymore today...